J. M

Paranormal Investigation

Hunting Ghosts Using Scientific Methods

Written By

J. Michael Atencia

The Paranormal Investigator

J. M. Atencio Publishing
1805 Nancy Drive
Ruston, LA 71270
318 436-1588
ISBN **1-4196-5276-1**
ISBN13 **9-781419-652769**

For information about discount purchases for bulk purchases for clubs or schools,
Call (318) 436-1588 or
email: mailto:postmaster@the-paranormal-investigator.com?subject=Book discounts
Cover Design by **Jeffrey M. Atencio**

I dedicate this book to:

My mom, **Jacquie Atencio**, who I lost Thanksgiving 2005. Beloved Mom and Nan, ***I FINALLY DID IT!*** Thanks for always being there when I needed you and for teaching me determination in the face of adversity, for always loving me, for the patience and encouragement you gave so willingly. I'll always love you and remember your smile and words of encouragement… as always, "***Thanks Mom, I won't say good bye, I'll just see you later***".

My wife **Laynie**, who single handedly kept the family together when I lost my mom and put up with me during my deep depression. For your moral, physical and emotional support which allowed me to recover and finish my degree and this book. Thanks for putting up with my temper and moodiness and staying on my tail feathers when I needed it most. Baby, you're the greatest!

My kids, **Donnie**, **Jacquie**, **Tony** and **Kathryn** who waited hand and foot on me while I was down, even when I yelled and grounded them for stupid stuff. I love you guys…

And finally, all my English teachers at Coronado Junior High and High School, Bossier Parish Community College (Ms. French) and Dr. Eddie Blick and Dr. Reggie Owens at Louisiana Tech University in the Journalism department for teaching me about journalism and Dr Stokley, my student advisor and friend and my dear friend Dr John Strait for making geography fun, now I can find ghosts anywhere and do a population migration census. John Rieck and Kristin Loke at Booksurge.com

I cannot say thank you enough, I certainly know I've been a real pain in the tail feathers. I can never repay you for all you have given me, **GOD Bless You Guys,**

J. Michael (Mike) Atencio

Table of Contents

Introduction

A few notes about this book. There are links to video clips in the first section about ghosts. It has corresponding video explaining the different types of ghosts. Look for the clapper board icon.

Visit this link for the video downloads:
http://www.the-paranormal-investigator.com/images/video

Ghosts are real. After you read ***Paranormal Investigation***, you will be able to go out and hunt ghosts; start a club and or take the wife/ hubby and kids out and go on a ghost hunting expedition across the USA or where-ever you live.

Whether you want to learn about ghosts, research paranormal activity or clear locations of spirits, I have written this book to assist you every step of the way. I have been a ghost hunter since I was 12 and became a researcher by the time I was 18. At 46, I have a lot of experience under my belt. I have taught friends, my boy scouts when I was a scoutmaster and my family now that they are old enough.

I was a police officer, a correctional officer where I was trained as an investigator and a as sergeant for Louisiana (Angola) State Prison. I have a B.A. in sociology, a minor in geography and in journalism.

I wanted to share my experience with more people because I really enjoy ghost hunting and it's starting to gain acceptance with people of all backgrounds and experience.

With this guide, I will share my knowledge and experience about investigation applied to this scientific field.

We begin by understanding what a ghost is. By defining what spirits are, we can be better prepared to encounter them and communicate if possible. If your afraid of bumping into an unseen person, remember this, most ghosts aren't the violent type you see in the movies. They were people before they died and they have no concept of time anymore. They really aren't interested in causing harm. Usually, they just want to be left alone or they like to play tricks on you. Occasionally, they may touch you or make a sound to let you know they are there with you. ***Remember this, FEAR feeds their energy, the more scared you are the more power and strength they get.***

You must believe that you have the power to really control them. That confidence gives you immense strength spiritually and psychically.

If you feel scared enough to run, **don't**. Walk quickly to an exit and recover. You can always return later.

I have included a basic list of the tools needed for ghost hunting but I want to emphasize that all you need is a flashlight, compass and a watch, a logbook, a pen to keep notes and lots of patience. Everything else you can purchase as you go. I have good equipment that is ten years old and I wouldn't trade it for anything.

Common sense and listening to that little voice in your head are the greatest tools of all and they don't cost a thing.

After you learn about the equipment, it will be time to learn about investigating.

There are many approaches to investigation. You can be a simple observer or a independent researcher, join or start a ghost hunter club or become a researcher and collect data about specific locations. Whatever your choice you can rest assured that there is plenty of information for it and I'm always available by email to answer questions. It may take time to answer your questions but I promise you'll have an answer ASAP.

I admit, the spirit realm may be far reaching for scientist and observer alike but rest assured, you stand in the ranks of many great people. Einstein, DaVinci, Ben Franklin, Alexander Graham Bell and many others all began their journey of exploring and explaining the possibility of an afterlife from right where you are… at the beginning.

Part One
What Is A Ghost?

What Is A Ghost?

A ghost is a disembodied spirit trapped by accident or intentional means between here and the spectral or astral plane. Contact is possible because the time-space continuum may have points in which they fold over. At these points is when we visually see the apparition. There seems to be a connection between weather, geo-magnetism and electricity. Suns spots may even play a role in contact.

There are many types of ghosts / spirits and I will go into some detail in Part One describing what they are and their differences.

There seems to be many ways ghosts can be in our world. Some are Earth bound, some stay by choice, some are here to finish incomplete business and some are afraid to cross over. Then you have your imprints which are a spectral recording of an event in a particular location.

Determining what type of haunting and ghost you have makes your job easier. Each have specific criteria you will apply to your investigation.

Paranormal Terminology

I originally decided to place these definitions at the end of the book in a glossary but it is more important to know these right from the beginning. There are many types of phenomena and as many terms. I will refer to these through-out the book so you should at least glance at them.

Apparition - a ghost you can see.
Anniversary Imprint - these imprints usually manifest around the same time each year.
Astral Projection - intentionally having an out-of-body experience.
Aura - a field of energy that emanates from matter. Some claim to see it as various colors. It is especially prominent around living things similar to a "force field".
Automatic Writing - subconsciously expressing thoughts or influences by doodling or writing.
Chi -Asian term for the "life force". In

Asian cultures it is the biological energy that is inhaled and can be manipulated for specific purposes (also called ki).

Clairvoyance - the ability to obtain knowledge based on unexplainable intuition, vision, or various psychic senses.

Clearing - removing ghostly activity from a location.

Cold Spots - cold patches of air in a warm room where two different temperatures can be obtained within feet or inches of each other. Believed to be ghosts trying to materialize.

Doppelganger - a ghost that looks identical to a living person but behaves differently. These are ghosts of the present.

Double - a ghost that looks and behaves identically to a living person.

Dowsing - using and interpreting the motions of instruments to obtain information (also called divining).

Ectoplasm - any type of physical substance accompanying a spirit's materialization. This is the goo (think ***Ghostbusters*** movie) on the actors but can also be a different material.

Electromagnetic Energy - a mixture of electrical charges and magnetic fields that bind nature.

Entity - an interactive ghost. It's conscience of its existence and of the living people around it.

ESP - acronym for "extra-sensory perception"; obtaining information from a source other than the five physical senses (i.e., sight, smell, hearing, taste, touch). Can be called intuition.

Etheric Body - a layer of the physical body composed entirely of energy. Mimicking the actual body.

EVP - acronym for "electronic voice phenomena". Can be heard in conjunction with white noise for capturing ghostly sounds and/or words on a tape or digital recording.

Exorcism - clearing a person or a location of evil spirits. Done by using religious rituals.

Ghost - a paranormal event appearing to be conscience of its environment. See Entity.
Ghost Hunter - one who seeks to experience and document ghostly activity.
Harbinger - a ghost of the future that brings warning of impending events.
Haunted Location - an area where ghostly activity occurs regularly, especially for more than a year. Some researchers refer to locations plagued by imprints only as "haunted."
Hot Spot - a site within a haunted location where activity is prominent and/or energy fields are focused. Can also is similar to a cold spot except air is noticeably warmer than surrounding air.
Imprint - ghostly activity that appears non conscious. This type of activity repeats.
Investigator – A person that researches and tries to explain using scientific means; paranormal activity.
Ion - an electrically charged atom or molecule.
Magnetosphere - the magnetic field

surrounding the earth.
Materialization - the process by which a spirit creates itself physically.
Medium - A person who claims to possess an ability to communicate with spirits through voices, symbols or pictures.
Natural - phenomena that appears ghostly but in fact is created by some scientifically unknown property of the present nature. A recurring banging turns out to be new plumbing pipes vibrating from air. Also known as air hammering. Sounds frightening but it's nothing paranormal.
Necromancy - resurrection and interaction with the dead, particularly for the purpose of communication or control.
Original Momentum - telekinetic force or energy necessary to create motion.
Out-of-Body Experience - when one's consciousness exits the restrictions of the physical body (also called OBE).
Paranormal Research - the study of phenomena currently considered unexplainable by science. Events or

phenomena can not be reproduced to the satisfaction of the scientific community currently.

PK - acronym for "psychokineses" (see Telekinesis).

Poltergeist - German for "noisy ghost"; an entity or energy that displays sensational interaction with the physical environment, and manifests usually when a specific individual or individuals are present. Mostly around pre-teen or teen girls with emotion instability.

Portal - a theoretical doorway of energy, through which spirits may be able to enter or exit a location.

Possession - the act of being physically or mentally controlled by spiritual forces, usually negative forces. (Think of ***The Exorcist***).

Precognition - sensing, seeing or knowing activity received from the future using ESP.

Premonition - a psychically created awareness of future events, often with a negative outcome.

Primary Readings - the beginning measurements of taken at a haunted location, used for establishing an investigation's investigation starting point. Usually done with an EMF meter and thermometer.

Psi - a general, all-encompassing term for "psychic phenomena.

Psionics - the use of physical tools to assist in accessing or interpreting one's ESP.

Psychic - phenomena rooted in ESP and spiritualism; also, a person gifted with ESP.

Retro-cognition - seeing or knowing activity from the past using ESP.

Revenant - an entity that comes back a few times after death. These spirits usually come back to take care of unfinished business.

Séance - a ritual held to communicate with spirits of the dead.

Spiritualism - belief in a spiritual world and/or the ability to communicate with spirits.

Synchronicity - the product of numerous, seemingly unrelated variables joining to create a common event or remarkable "coincidence."

Telepathy - the process by which a mind can communicate directly with another without using normal, physical interaction or ordinary sensory perception.

Telekinesis - the ability to control one's physical environment without using physical manipulation or force (also known as psychokineses, TK, or PK).Moving things with the mind.

Warp - a location where the known laws of physics do not apply and space/time may be distorted.

Now that you have reviewed the terminology of the trade; it's time to discuss what a ghost is.

Many believe that when they die they go to heaven. Ask then what heaven is and you'll get a multitude of answers. Much depends upon which religion the person follows and the amount of faith they have in that religion's god. Generally speaking; if you are a good person and have lived a good honest and respectful life; you go to heaven. If not then chances are you'll end up in Hell or some form of purgatory.

I subscribe to the belief that when I die; my soul leaves the body and goes to my conception of heaven. I cross over into the light. I also believe that I will have a chance to resolve any unfinished business if it's something that must be finished. I will say good-bye to my loved ones when they are ready for me to do so and then I'll complete my final journey to happiness and peace; where I'll meet up with my mom and people that arrived before me. To me, that is heaven.

Your idea may be different. That's fine. I believe and theorize that whatever your lifelong belief, that's what your heaven will be.

Mediums say that some ghosts are in visitation. Some are inherently bad and have been grounded to a location and some are confused and don't realize they have passed and get stuck between here and the afterlife. Some don't leave because they simply don't want to and finally; some stay intentionally because they lived an evil life and are afraid of what is waiting for them. They don't want to be judged for their sins and they are afraid to go into the light.

Knowing this helps you the investigator in many ways. When you research a location; you'll have an idea of who you may be dealing with. Remember; ghosts were people too. Knowing the history and circumstances helps you understand the type of ghost or entity you'll be dealing with. Usually; tragic deaths or murders tend to leave a strong impression on the environment. Not surprisingly; a strong love of family or home leaves just as strong impression. These are strong and explosive emotional events. Hauntings appear to bloom out of this emotional turbulence.

We will go into specific detail hereafter what is a general consensus of paranormal investigators of what a ghost is. There are several basic distinctions but it is important to remember that people have different personalities and so ghosts will have different personalities as well. Some are endearing and kind hearted and some are pranksters. You may find some don't want to be bothered and others want to talk your ear off through EVP communicating or knocking etc. Some like to materialize and float and others like to move or hide items.

So what is a ghost? A ghost is a disembodied spirit with as varied a personality as anyone in the living world. I suggest that you approach each investigation like you were a social worker and you want to understand the spirits situation so you can help that spirit if necessary or wanted. How would you want to be treated if you were a spirit?

Entities

There are different types of entities including humans; animals and even dinosaurs. Any living being can become an entity after dying. According to ***The Field Guide to Ghosts and Other Apparitions***, by Hillary Evans and Patrick Huyghe, they organized ghosts into past, present, and future categories. Here's a breakdown of each group.

Ghost from the Past:

Ghosts from the past are the ones we're all most familiar with. The term for an entity to come back only a few times after death is *revenant.* This is usually a spirit that is tying up loose ends or has some unfinished business to take care of before leaving this plane. For whatever reason, the spirits procrastinate before crossing.

Ghosts of the Present:

Although extremely rare; there have been reported cases of ghosts of living people. Out of body experiences or O. B.

E. are situations where the spirit leaves the body briefly. These people claim that they have left their body after dying in we're able to observe their body during surgery or something similar. Others claim they can do this and control it. This is called astral projection. Much like a superhero, they fly all around when their bodies are asleep seeming completely weightless. They claim the skies are filled with thousands of astral travelers.

Ghosts of the Future:

Ghosts of the future much like ghosts of the present only appear once or twice. In these types of situations the person is known as a "harbinger" or bringing a message to inform the observer or future events. This is usually a warning. Harbingers can also prevent a death by warning of a life-threatening situation about to occur. Soldiers in combat, children whose parents visit them at night while sleeping and so forth have reported this type of harbinger.

How to Identify an Entity:

- ***These are conscious beings.***
- ***In most cases these are people or animals who've died a tragic death.***
- ***Batteries and flashlights, cameras, video cameras or other electrical storage can be drained to help the spirit manifest.***
- ***Can come from the past, present, or future.***
- ***The experience can be objective, subjective, or both.***

Visit this link for video downloads:
http://www.the-paranormal-investigator.com/images/video

Right-click on each video link and select "save target as".
It will ask where you want to save the video and you save where you want to on your computer.
You'll need Windows Media Player to

view the videos and speakers to hear the videos.

Imprints

Imprints are different from entities in that they are inanimate objects. A good example is *The Flying Dutchman.* For nearly 200 years everyone that sees this ship has misfortune. *The Flying Dutchman* usually appears the storms are in the vicinity. It emerges from a mist, it flows across the dark sea, then; in a glowing spectral luminescence; it continues until out of sight. Upon seeing the ship; the person or persons should dock their ship immediately or suffer the consequences.

Another way to look at imprints is to think of a videotape or tape recording that's played over and over and over again. A tragic event like a murder or accidental death can imprint itself in a location and under certain circumstances replay itself. Many researchers will call this a haunting however; I feel entities ***or*** imprints fit the description of a haunting. One thing to keep in mind, unlike a movie clip or a video which is two-dimensional, an imprint shows in three dimensions. It can be seen from any angle and appear totally realistic in every way except the people involved are totally oblivious to the living people in the room or location.

Another form of imprint is scent. Frequently the smell of flowers, roses in particular, and cigar smoke are reported. These are imprints; also; they can be from an entity. Again, the best way to identify if it is an imprint or an entity is that the scent will move around with an entity where as an imprint smell will stay in one location.

Sound is also an imprint. Again, if the sound is in one place repeatedly then it most likely is an imprint if the sound tends to move around the most likely is an entity war and natural. For more information about naturals click on the natural link.

Poltergeists

The term poltergeist comes from the German term meaning noisy ghost. It's been found that a poltergeist will usually center around one person. Often it's an adolescent; highly emotional female but anyone can be an agent. In the past, poltergeists were thought to be mischievous spirits that were extremely interactive with their environment. These "ghosts" would move or throw things, pop bottle tops, make lights flicker off and on; make loud noises and so forth. This type of activity seemed to be very cruel and really enjoyed doing malevolent. Unlike entities or imprints, poltergeist phenomena tend to last just a short time and in the old days it was believed the devil was involved in some way.

Young females tend to have a lot of emotional energy and it seems that poltergeist activity can feed off of this energy and then use the energy to throw things etc. Another alternative for this type of phenomena is telekinesis or moving things with the mind.

Unlike a conventional haunting, poltergeist activity must have a person at the center of the phenomena. Another thing about this type of activity is it only usually last for a few months, is primarily around young women, (young men can be affected also but very rarely) who are experiencing hormonal changes. In rare circumstances it's been known to spontaneously develop throughout a lifetime completely without warning.

People with poltergeist phenomena occurring around them would tend to believe the entity is following them where ever they go. Often it's very hard to discern between an entity and poltergeist activity if phenomena occurs when that person is not around. It is possible for poltergeist activity to be occurring and for there to be an entity present in the same location at the same time. Unfortunately, there is no way to make the poltergeist activity stop except through professional counseling. If the activity is occurring because of emotional or telekinetic outbursts; then it is presumable that once this disruption of the emotion is resolved so will the poltergeist activity.

Naturals

Naturals are phenomena that appear to be ghostly in nature are actually very normal things appearing to be supernatural or paranormal. A good example of a natural event is a banging on the walls. Let's say for example that is banging occurs at night about the same time shortly after everyone goes to bed. Night after night the banging is heard but after a moment or two, it stops making it impossible to identify where the sound is coming from. It is later found that his paranormal activity was nothing more than the sprinklers coming on at night which caused the pipes to knock under the house for just a moment.

Another type of natural phenomena is one where the occupants of the home feel dizzy or ill, see things or shadows and generally have a sense of dread. This phenomena will occur in the same location at all hours of the day or night. A sense that the hair is rising on your neck can accompany this or the feeling of being touched by an unseen force.

These are all symptomatic of electro magnetic field poisoning. This can occur when electrical devices or wiring are not shielded properly allowing this unharnessed energy to emit in one area in such a high quantity that anyone entering this location will be subject to these symptoms.

The use of an E.M.F. meter or electro-magnetic field meter will detect the strong fields emitted by devices or wiring. The good thing about this type of phenomena is that it can easily be remedied by an electrician shielding your wiring.

The difference between this type of phenomena and an entity or haunting is that the E.M.F. meter will have a higher reading in one location and during a haunting the media will show variable readings that tend to move from location to location. In any situation, is imperative that repairs be made immediately it is found to be problems with your wiring or a device emitting a strong and dangerous field.

Let me share with you and experience I had personally. I was staying the night at a friends house sleeping on the couch. Sometime during the night I was awakened by a sound in the kitchen. I lay there on the couch terrified beyond comprehension. The blender kept turning itself off and on all my long. I pulled the covers over my head and hid. I eventually fell asleep after several hours and when I woke up, it was quiet. I told my mom and my friends about the experience and of course everyone thought I was nuts. I was there the rest of the day and the blender never turned on or off by itself again.

I was pretty upset about it because people thought I was making it up. A week or so later, our friend call to tell me the blender kept turning itself off and on and kept them up all night long. She took the blender to a shop and it was discovered that the switch was faulty and had a short which caused it to turn off and on randomly. It still scared me pretty bad and then I felt foolish over something so simple as a shorted switch.

Naturals can also cross the line between normal things and paranormal things. Some paranormal activity can be both natural and be explained scientifically and yet parts can't be explained at all.

An example: Say you were watching a door that seems to close itself for no reason at all. You investigate and find the door will close on its own because it's hung on a tilted door frame. OK, this is easy to explain so far. Now you notice that it can't close completely because the floor stops the door at a certain place.

After hours of investigating you stop for the night and check the door. You find the door has closed as predicted except it has also latched itself. How did it do that? That's the part that's paranormal activity. It's both normal and paranormal.

Part Two
Who Hunt Ghosts?

Why Hunt Ghosts?

The first thing to remember is that the search and belief in an afterlife has been the driving force of humanity from the beginning of time. Every religion is based in some way or another on there being an afterlife. It only makes sense that we would look for scientific proof of it. The easiest way to do this is through ghost hunting. Aside from enjoying the thrill of conquering your fear of things going bump in the night; it sure makes all that hard work of following the golden rule worth the effort.

Imagine what it would be like if people didn't believe in an afterlife. There wouldn't be any laws or moral values in place because it really wouldn't matter.

That thought scares me just thinking about it. More so; Darwinian's believe we are created, we live and we die. End of story. OK, so why do we have paranormal phenomena? Darwin theorized evolution but he was a religious man. He was using common logic to explain the unexplainable.

Technically, Darwin was a paranormal investigator. So was Albert Einstein, Ben Franklin, Thomas Jefferson, Alexander Graham Bell, Marie Curie, Louis Pasture and Leonardo DaVinci. All searched for an answer to a question they had.

Now you can join these investigators by becoming an investigator yourself. You only need a strong determination to find the truth. A willingness to devise a theory and then set out to prove it using scientific procedure. Most of all, common sense and logic are the best tools you can use.

You can choose to experience ghostly phenomena. You do not have to do scientific research. I can tell you this, once an entity touches you; you never forget it and you'll want to know more. That is how most of us get into this work.

If you have dreams that come true

(deja vu), or had a visit from a lost loved one, then you will be looking for answers the rest of your life.

On a more earthly side; we all want to experience something paranormal or we wouldn't be doing this.

Imagine what it would be like if a group of school kids could go to Kitty Hawk, North Carolina and watch the first flight of man as it actually happened! They could just put on a pair of goggles and a head-set; dial into the frequency and observe the flight.

In theory, it should be possible. We just have to develop equipment to find that frequency. Until then; all we can do is observe and report when we catch a glimpse of these phenomena.

About Ghost Hunting

Ghost hunting should be a fun experience. Like anything worth doing; ghost hunting requires work and patience. When or if you have children; expect for them to become bored or need to use the bathroom frequently. These are things to expect. Plan for them. All the best equipment isn't worth two cents if you have to leave every 15 minutes to find a potty for little Timmy. Same goes for people on your team with weak bladders or after you have been investigating for several hours in a location sans restrooms.

Porta-potties are a great investment! Food and water are also a BIG thing to plan for. If you are in a hot location; don't take caffeinated drinks. These dehydrate you quickly. Bring lots of chilled water and sports drinks too. If you're climbing stairs and doing lots of walking, comfortable shoes are a major thing to invest in. If it's cold, blankets and heavy windproof jackets are a good investment.

Cars will breakdown when you least expect it. Before going on a trip outside of your local area you should have a complete inspection made of fluids, belts, hoses, brakes and tires. Change the oil and get a tune-up. Wash and clean out the car.

Check your camping gear if you're planning to camp out. Air out sleeping bags and tents. Check to make sure the stakes and guide ropes are in useable shape. Get lantern fluid and extra propane for your stove. Get an inflatable mattress because your back will need it if you have been driving a lot. A cot is even better.

I suggest motels to campgrounds but if you are traveling cross country with kids, campgrounds are much cheaper and fun. Roasting marshmallows and wieners over a campfire while telling scary stories is a great way to start off a ghost hunting expedition. With my family; we do the motels and go to the beach for the wieners and marshmallows. Then ghost hunt between amusement parks and zoo visits.

That way everyone is happy. My wife isn't an outdoors type. Bathrooms and comfortable beds is her preference and I have to admit; I've found it great to sleep in a comfortable bed after a hot shower and a great meal. Old age does that to you – or at least me.

Most importantly is to leave the stress at home. If things don't go as you expected; let it go and adapt to the new situation. It may be a spirit guiding you to an adventure you never considered or thought you would have. Some of the greatest times I have ever had happened when my plans went to hell and I just with it. It always works out and now I welcome these little changes.

My philosophy is: if it weren't for all the bad or inconvenient stuff happening, there'd be nothing good to look forward to. It would get pretty boring real quick.

What causes a ghost to manifest itself?

A lot of things can cause this to

happen. Atmosphere and moisture, static electricity charges, remodeling a building or home are the most frequent causes. Remodeling is the one I have seen most of the time. Spirits hate having their home changed for some reason.

What To Look For

Now that we know what ghosts are and we're on our investigation what do we look for. The answer is anything paranormal. Cold spots are a give away but they don't always occur. Scents, bangs, bumping or knocking sounds are indicators too. The best thing to use is your intuition. If the hair on your neck starts rising or you sense something different in the air (literally) then you may be experiencing ghostly activity… or a draft.

Equipment with batteries tend to drain quickly when a spirit is trying to materialize. That's a dead give away (no pun intended).

Outdoors, all of the normal sounds will cease. It will become absolutely quiet with none of the night creatures making a sound. Of course, if you just arrived, they may be silent because of your arrival. If this is the case, stay quiet for a few minutes and the crickets will start talking again.

Set up trigger objects. Trigger

objects are items that may be moved. As an example, take a coin and put it on a piece of paper, draw a circle around it and ask the spirits to move the coin. Leave the room sealed and secured and go on with you investigation. At the end, come back and see if the coin has been moved. If it has and the room was locked; you had a visitor.

A sense of foreboding or dread. Depression or the feeling of being watched. Sometimes a nudge or that someone is standing by your bed staring at you. These are things you intuitively pick up on without realizing it.

Mirrors when facing each other are believed to amplify energy and create a portal as is running water.

Remodeling a house will cause activity also because you are disturbing familiar surroundings of someone that has passed on.

It's very common for activity to increase during autumn and winter when it's cold and low humidity.

When You Find Them

Now that you have an idea of what to look for what do you do when you find them? That's easy, observe and report and try to catch them (on tape or film) and communicate with them. Let's talk about these things.

Usually, but not always, you'll experience hot or cold spots, the hair on your neck or arms may rise and you could hear noises like footsteps, banging and knocking or something like a door opening and closing. These are BIG signs that you are being graced with the presence of a spirit or disembodied soul. They aren't as obvious most of the time unfortunately. Ghosts are VERY shy. Communication is the easiest way to make contact.

There are many ways to communicate with spirits and we can explore them here.

Psychics and Mediums

These folks can communicate in a number of ways. Mediums communicate through symbols, pictures, letters or numbers etc. They then have to make sense out of the information a spirit is giving them. They don't get a booming voice saying "*Tell Auntie Gertrude I'm doing great!*" It doesn't work that way. They are given a series of symbols and the have to piece together the message and then translate it for us. Psychics can have many varied talents, visual, audible by smell or intuition. Many combine the term psychic and medium but they are different to a point. Someone that is channeling information through spirit contact is a medium. Psychics can touch and get impressions of events and tell you what has happened in the past or who was present and so forth. Mediums will charge in most cases for their talents and that's fine. I feel as long as they are using their gift for good then it's not a problem to make a living from their gift.

Think of it this way, you're a construction worker and you build a house for someone, do you do it for free? That's the point, nobody can afford to work for free. I don't and you shouldn't either. Of course, they need to earn their money which means they must produce with-out any clues or hints at all. A reputable medium won't mind and will welcome the opportunity. Psychics are and should be treated the same.

The following are unscientific ways of communicating with the other side. I don't use any of these in my investigations. How can I collect empirical data using these methods? I can't and you can't either. They are fun "party tricks" but that's about it. Try them out for fun and gaining knowledge.

I use psychics and mediums. They are a tool to help me understand what or who I'm dealing with but again; they don't provide empirical evidence. If you are doing scientific investigation leave these communication devices behind and focus on real evidence collection. Include

mediums and psychics but understand that they can make mistakes or interpret information incorrectly.

Ouija Boards

These boards are supposed to be a portal to communicate with an entity. I have mixed feelings about them. They require two people (usually a man and a woman) to lightly touch the pointer and let the spirit tell them answers to their questions. I understand that one person can use the board but I never have tried this.
A common influence we need to discuss is "**automatism**". Automatism is your pulse influencing where the pointer may go. Unintentionally, you want the guy or girl to like you so your mind pushes the pointer in that direction. It's quite simple really.
As for whether they really work or not is debatable. Some believe they can open doors to entities and place the user in mortal danger of being possessed; others believe that they release entities into this world. I don't think there is any proof that these boards work. I've never seen any evidence. I don't waste my time on these things but if you want to, go for it.

You can make your own board. Write the alphabet on a board with enough room for a small glass to cover a single letter at a time. The board must be smooth enough for a glass to slide easily. Put **Yes** and **No** in corners and **Maybe** and **I Don't Know** somewhere they will fit.

Using yourself and a friend while touching the glass lightly, ask a question and see what answer you get.

Dowsing Rods

The rods are designed to locate energy. When an energy vortex is reached, the rods will cross. The stronger the field; the more the rods should cross. These are two nine inch metal rods of equal length bent in a 90 degree angle to form handles. The handles are placed in tubes or plastic straws so they won't be hindered.

I made mine from a brass rod. You can make it from a metal coat hanger. Cut the hook part off then cut two pieces about a foot long from the remaining wire. Bend one end so it fits in your hand (about 4 inches)

Rods are set so they are balanced and can swing freely. User should not tilt the rods so they'll cross. They are supposed to cross on their own. It takes practice to walk and balance the rods and like everything else an investment of time.

These have never worked for me – ever. Again; try them out for fun. Maybe you're sensitive enough to pick up the "vibes".

Automatic Writing

This is doodling under the guidance of a spirit and something legible appears in the doodling. Automatism applies here too. Subconsciously, the mind writes what it is thinking about even though the writer may not be aware of it. I figure you can do it if you want. It can't hurt, right.

Séances

These consist of a group of people sitting or standing around a table and summoning spirits for communication. The people hold each other's hands and say "We call upon the spirits of this location to communicate with us. Please let us know that you are here. We will not hurt you. Mrs. XYZ is here and wishes to speak with Mr. XYZ." Then everyone waits and continues holding hands until an answer is given. Ask the spirit to knock twice for yes and once for no. Hands must stay clasped.

If nothing else, this is fun to try. The only other place you can hold hands is at Thanksgiving (American holiday) dinner. If you do try this, make sure you have a video camera watching under the table, in case someone is being funny and moves the table with their knees or feet. There's always a comedian around.

I have been to a few of these but I have absolutely no faith in these at all.

Table Tipping

Sit or stand in a circle around a table and lightly touch the table top with the tips of your fingers. Say "I ask the spirits of this location to show they are here by tipping or moving this table" and watch for movement.

Both séances and table tipping can be easily faked so they should be performed with people you absolutely trust.

The Pendulum

A pendulum consists of a string with a weight at the bottom. It's held over a piece of paper with what looks like a target and crosshairs on it and questions are asked. If it goes left and right the answer is yes and up and down the answer is no. It can be clockwise for yes and counter-clockwise for no also.

Again automatism applies here. You can make one from a piece of string or fishing line and a small fishing weight with a pointed end at the bottom. Take a piece of paper and draw a circle about two inches in diameter. Then draw a cross in the center. See my example below. I sell these on my website if your interested.

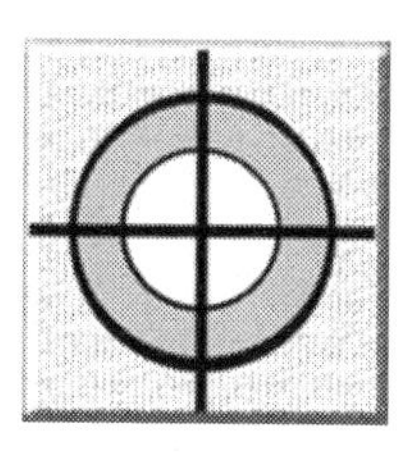

Make the string about a foot long and tie the weight to it. Position your arm over the target and be very still. The pendulum will begin to move in a moment or two.

Again, try it out. If you get the same answer as two or three others in your group, you may be on to something.

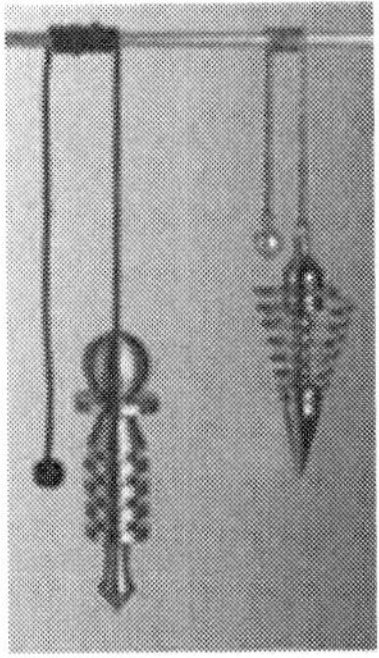

Examples of pendulums

PART THREE
THE
INVEST GAT ON

Before we go further; I want to take a moment and comment about buying lots of equipment. Don't do it until you feel ready to and then only if you have the money to use on toys you'll use occasionally.

This is a hobby. I buy a piece here and the wife and kids get me a piece there on days like father's day; birthdays or Christmas. Get only what you need; not what you want.

If you do this hobby full-time then the equipment is much cheaper than skiing equipment.

Tools of the Trade

There are many tools that are available to investigators but you don't have to spend thousands of dollars to get the job done. I'll include a list of items that you need and an extended list of items that you can purchase as you get into investigating seriously. I don't recommend any brand over any other brand. Tools are a personal thing and everyone has their pet peeves.

I have included this link to a free program for ghost hunters. It is a utility program that allows you to track your cases, equipment and do other stuff.

Click on this link:
http://www.download.com/Ghost-Tech-Paranormal-Investigator/3000-2054_4-10556320.html?tag=pdp_prod

This hobby should be as fun as possible. Loading up with tons of equipment and wires running every which way is not my idea of fun and it shouldn't be yours either. This hobby is a lesson in patience. Everything happens in its own moment. Here is the basic list:

Flashlights

A flashlight is the most invaluable tool you can have. Actually, several flashlights are what you need. I use a Mini Maglite with the optional red lens kit. I also have the nylon belt holder for it and over the last ten years; it has provided me with quality and service every day. Being a former cop; I can't begin to tell you the value of a good light at night. The red lens makes it easy for my eyes to adjust to the darkness whereas the white light hurts my eyes and makes me see spots (think camera flash here) everywhere for five or ten minutes. Late at night and being tired, those spots can look like "ghosts" or shadow people.

I also recommend a couple of large flashlights that have the water resistant feature. You can also get a hands free light that mounts on a hat or your head like a miner's light. These are ideal for keeping your hands available for other things like writing in a log book or holding a EMF

meter.
I use a Mini-Maglite.
It can be used as a spot,
flood or candle. It
even has an accessory
pack as shown below.

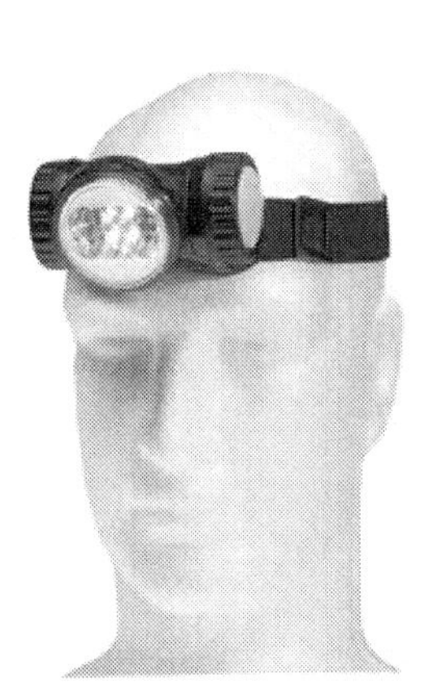

Batteries

ALWAYS carry extra batteries for every item that uses batteries. Why? **Because ghosts love to use the energy supplied by batteries to materialize.** They don't always do this but they will if you don't have back-up batteries. Think of **Murphy's Law.**

If anything can go wrong, it will. I'm a firm believer in Mr. Murphy and his unique sense of humor. Keep a recharger for your batteries too. If you have teens, these pay for themselves almost immediately. Recharging AA and AAA batteries is much cheaper in the long run. Get the large 9 volt batteries for the big flashlights if you use that type of light and an extra battery for your camcorder. These batteries run down quick. My camcorder has a recharger but it takes a while for it to charge the battery, like an hour or so. Extra batteries save me a lot of headaches.

There is another thing to be mindful of. Sometimes equipment may operate without batteries in them. During hauntings; toys or equipment may inexplicably operate and there is no battery in the battery compartment. I've heard many stories about "possessed" toys. In the end, the toys had to be destroyed to terminate their operation. Don't be alarmed if this happens. Have you ever heard of someone being killed or hurt by their toys? It's just another paranormal experience to explore and record.

Watch

You really need a watch to record events. Every scientist will tell you time recording is extremely important. I recommend a water resistant type with the blue night-light feature. I've even had to use this when we lost all power during a storm and had no lighting at all. I pushed the button and all the kids found me. They could see the blue light. It gave them and me a sense of security. The time is important also because it gives you a reference point to compare with your phenomena.

Any standard watch with a night visibility feature works well for ghost hunting.

Compass

The beginner will find using a compass very easy. Normally a compass points to magnetic north. If you have electromagnetic field fluctuations; the compass will be mis-directed because the magnetic field is being disturbed. This is what an EMF meter does; detect changes in the fields. All you need is a steady hand and patience watching the compass for movement. If it moves; there is a possible ghost nearby.

A compass is also useful for orientation at hauntings. There are times when you will be outside and it will be helpful if you know how to use the compass to figure out where you are standing from landmarks such as trees or gates at cemeteries etc.

A good compass is about $20.00 I prefer the type with the fold up cover that has a wire in the center for sighting on maps. A camping supply store, army surplus or Wal-Mart store will have these at affordable prices. You want to make sure it has a liquid dial. This means it has liquid inside a sealed dial. The dial has marks from 0 to 360 degrees. That dial turns and the needle floats inside. For the kids, use a cheaper compass. They tend to lose stuff or break items and it doesn't make sense to buy expensive compasses here. Get yourself the good one.

These are the basic tools you'll need. From here the tools are more advanced.

Electromagnetic Field Meter

An EMF meter is invaluable to ghost hunters. It tells you when the electromagnetic field is disrupted. It's believed that these disruptions occur during paranormal activity. It also distinguishes between harmful EMF's and ghostly activity. High readings are usually unshielded wiring or electrical panels. In turn, this is very dangerous to anyone exposed for long periods of time.

Residents report headaches, feeling dizzy or ill, and a sense of dread and possibly hallucinate; seeing shadow people and ghosts. Your meter will help you clarify what's going on.

Be sure to check high near the ceiling and low near floors and receptacles, bathroom fans, appliances and vent-a-hood or around electrical panels. (Don't touch wiring or panels!)

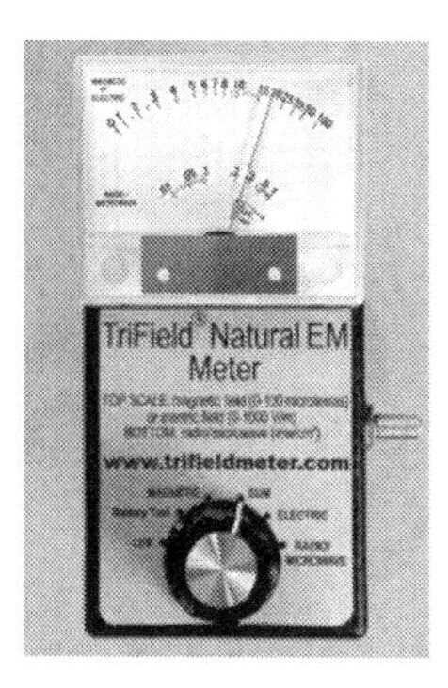

Two examples of EMF detectors
Visit <u>Supermeters</u> to purchase a meter.

Still Camera 35mm

A camera is an excellent tool for documentation. If you have a cold spot, start taking pictures.

I need to caution you that you'll need to load your camera with film for very low light conditions. That helps to make better pictures but it's something you have to experiment with. I have an expensive 35mm with auto forward winder and it's capable of 1/2000th of a second exposure settings. I picked it up in a pawn shop for about $50.00. Nobody uses film cameras anymore so I got a camera worth several hundred dollars.

Film speeds of 100, 200 and 400 are what you should keep with you. 100 and 200 are for low light and 400 is for normal use. The higher the number the more action you can capture under high light conditions.

Get a case for film and camera mounts and straps. Get a tripod for long exposure settings. It can be used for objects that move and catching that movement during time lapse photography.

Still Instant Picture

(Polaroid One Step®)

I use the camera named Polaroid One Step as an example here but any one step instant camera will work. These are great for seeing what you just caught; if anything; immediately. Imagine that you catch a ghost and are able to see it right away! With regular film, you have to have it developed. Even one hour places are inconvenient here.

I've read that the Polaroid film catches apparitions better than other films but I can't substantiate that claim.

These are still sold at many stores including Wal-Mart for about $35.00.

Still Camera Digital

These are the best of both worlds. Instant pictures and no film. Unless you have extra memory cards; expect to download a lot and reviewing pictures will use batteries up quickly. There is also lighting conditions to contend with. Consult your owners manual for work in low-light conditions. Digital cameras have come down in price.

Example of a digital camera

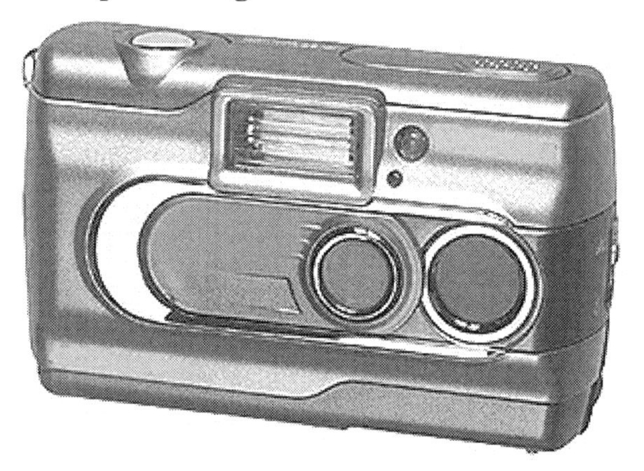

VHS Video Camera

I will use VHS here but it applies to Super VHS, 8mm, Beta cam and VHS-C.

I was a news cameraman for the local affiliate NBC and I used and edited many a tape. Here is the 411 on using these cameras. The cheaper the camera the better the low-lux (light) capability. Why? A standard VHS camera has only one chip. These are the eyes of video cameras. They don't have real good reproductive capability. In other words; if you try to re-record from a VHS tape; the quality will degrade. This is also known as a second generation recording. What I do is download to my computer from my camera and it converts it to a digital recording. I'll go into what you need installed on your computer later.

The better the camera; the more sensitive the low-lux capability and the more light you need. What I mean by more light is better quality light such as halogen lighting instead of regular room lighting. In the situation of ghost hunting; zero-lux is best. Use only the highest quality tapes too. They are more expensive but in the end pay for themselves many times over.

Typical VHS Camera

Digital Video Camcorder

Digital cameras come in different sizes also. There's mini dv, 8mm etc. The rule still applies here. They need light from somewhere unless the camera has night-light built in. You've seen this when the picture has a greenish glow and the people in the shot have those alien looking eyes. That's called night vision. Many digital cameras don't have this feature. The manufacturers omit this to make a better picture. Instead you'll see back-light. It's not really the same but it helps some in low-light situations. Check your owners manual for specifics or before purchasing a camera.

A positive thing about these cameras is that can double as a digital still camera with much better results. Download the camera into your computer and you can grab a frame from your footage using the right software.

This is similar to the camera I have. It has a night-light feature.

Infra-red lighting

Infra-red lights are lights that the camera can see. The eyes can't see this type of light because it's too low on the spectrum. Cameras see fairly well in the dark with IR lighting. I'm not sure of the costs but they aren't too expensive. It really comes down to how much investigating in the dark you are going to do.

Infra-red light

**Most newer cameras have a filter built in to block infra-red photography because it distorts the color. You'll have to shop for this camera feature if you use infra-red lighting. Shop for the camera before purchasing the lighting.

Tape Recorder

Just a plain old cassette recorder with a plug-in type microphone jack for an external mic is all you need for collecting EVP's. Generally you'll need to have white noise in the room. White noise is like a radio or TV channel without a signal. Water running from a spout is a good background noise too.

Drawbacks are the constant tape changing and the risk of a tape getting tangled up. I know from past experiences. This seems to happen in the dark when you are unable to do anything about it.

A cassette tape recorder

Digital Recorder

Digitals are just like a cassette recorder except they don't require a cassette. They also have files to store your EVP's in. Drawbacks range in the batteries they use up and the accidental deletion that's possible in the dark. Don't ask how I found this out, just trust me.

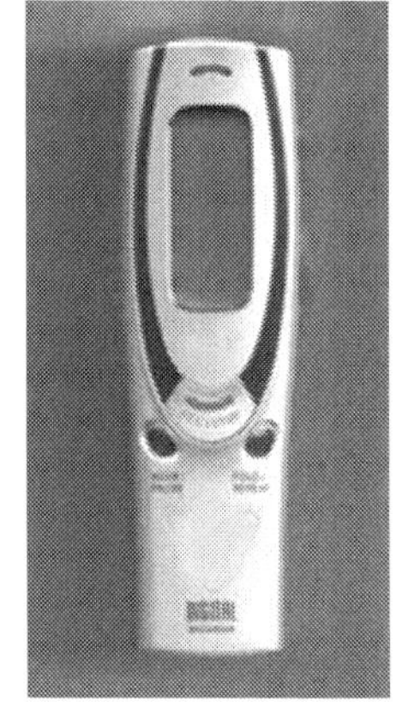

Examples of digital recorders.

Laptop with Speakers

This is an invaluable tool once you get into ghost hunting. If you have a desktop computer; don't worry. You can do everything on a desktop that you can do on a laptop and then some. You just don't have the convenience of a computer on site. I have two computers. A laptop and a desktop. I use the laptop for small projects and then transfer up to my desktop. I bought the laptop for $299.00 on EBay and it's about half the speed of my desktop. It's graphics card, hard drive and memory are smaller and it takes longer to start up. I use it for writing in my electronic journal and downloading and storing pictures from my digital cameras.

Speakers are definitely a must for any computer you own with a good sound card. I recommend this if you are going to collect and review EVP's. You can even use your computer to record EVP's and then play back the recordings.

A major warning here about

laptops. Laptops have a connector cord that plugs in. If it is damaged in the slightest way, your computer is toast. It costs nearly as much to repair as buying a new laptop computer.

Apple has a new laptop with a magnetic cord. If you trip over the cord, no worries mate, just put it back on. Dell (which I own) has a square plug with good support.

Sorry, I had this happen recently and it seems like they could have designed this much better. Now I have to buy a new laptop.

Desktop Computer

The desktop is my best tool for video and pictures. I run my website from it and edit video. I also listen to EVP's on it. I recommend investing in a good (expensive) graphics card and sound card. For software programs I have Pinnacle's Studio 9 Plus for video editing. I tried Adobe Premiere Pro and it was complicated to learn and use. The Studio was a better fit for me as I didn't need to spend months learning how to edit and create video clips. It is a drag and drop type of software and with-in an hour, your up and running – or editing tape.

This is where I do most of the work and then put it on a disc for the laptop and use the laptop for the reveal.

Again, don't spend money if you don't need to. Most times people have a computer and you don't need to bring one with you. Like I said, it's really for convenience.

A Digital Thermometer

This is a tool you really need. It reads temperature with a beam of infra-red light and can give fast and accurate readings. Hand-held with a big back-lit window for night-work.

Cost is about $30.00 to $80.00.

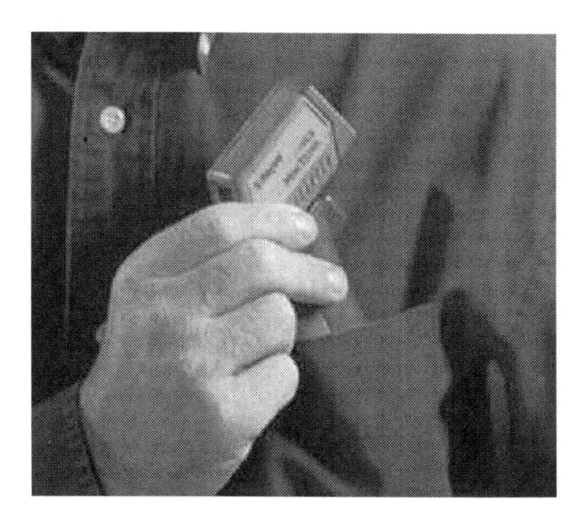

Motion Detectors

Hotels and public places have problems associated with them and motion detectors tend to keep folks honest. I use these in hallways or rooms with the doors closed and secured. If they get tripped (activated) the alarm alerts you and you know and you have activity in the room. They are cheap to buy and two or three work for most investigations.

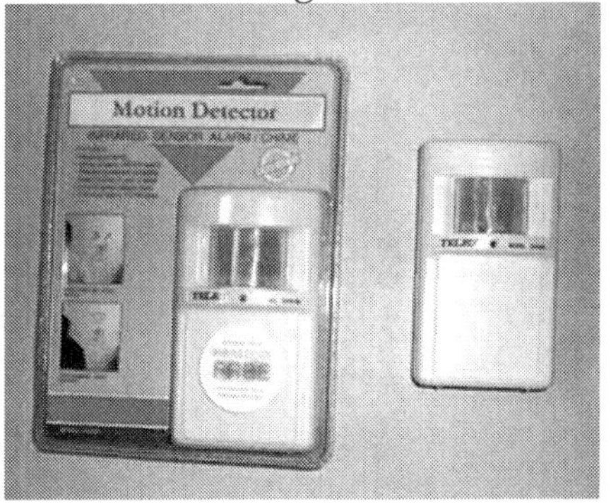

Walkie Talkies

The cheap ones won't work for investigations. You'll have to buy some quality walkie-talkies for about $40.00. Obviously the use of these goes without saying. These really come in handy so you team isn't yelling to each other.

I own the type that have 22 channels. I can select which channel I want depending on what I'm picking up (truckers etc.)

Portable Radio

I use it for white noise and for music or weather reports when needed. Dancing or singing too depending on how bored we get. Not really, but late at night in a room in never never land – it's nice to put a CD in and relax before bed. Some places have the most boring programs on TV at 4 AM.

Log books

I have included a link to a software program that has a logbook in it. I also use the old fashioned notebook that fits in your shirt pocket. I used these when I was a cop and just never got past keeping one with me. Any size will work but I prefer the small wire-bound type. A pen fits perfect in the wire part.

I'll discuss log books later in greater detail.

Client Info Sheets

You'll have to make one of these up with the pertinent questions and data. Name, date, phone number and address etc and an area asking what they are experiencing and if they believe in ghosts. You can download the free software to create a client sheet also.

Duct Tape

Used for taping stuff down like cables, wires and so forth. **Do not apply** to furniture, walls or flooring that could be damaged when removing the tape. I use it mostly on tripping hazards to tape cables down to carpeting. As long as you pull it up in a few hours and you're not in direct heat or sunlight, it comes off easily.

This stuff also works great for emergency ponchos and repairs on down jackets. It's embarrassing to have feathers going everywhere during an investigation.

Extension Cords

I never have enough of these. I ONLY use the kind with a circuit breaker built in for the first cord and connect a power strip w/ overload protection for my equipment. This protects against lightening strikes or power surges.

Use the heavier cord for construction and life will be great. Use a 12 or 14 gauge and you risk overheating it and starting a fire. A portable ceramic heater pulls a lot of power during a cold snap!

These more expensive cord with a breaker attached is a big life saver literally. Remember, if you're investigating a location in the dark, you might not see rain water seeping in. Mix that with electricity and someone is going to have a bad day

Baby Powder

I lightly sprinkle this on plastic trash bags taped together to see if footprints appear. It also makes it easy to see if fingerprints show up on smooth surfaces. Don't get carried away with the powder. Residents tend to get upset about dust everywhere.

Plastic Trash Bags

Great as emergency poncho's, for hauling trash and for catching footprints. Also good if you get wet and need to put those clothes in something other than on your car seats. Bigger bags (55 gallon) and the smaller ones are something you should always have on hand.

No-stick Cooking Spray

Again, like baby powder, this is used to see if prints appear on smooth or glass / mirrored objects. ***Make sure you bring window cleaner and paper towels to remove the spray at the end of your investigation.***

Water

Always have a gallon for each member of your team or two gallons during very hot weather. You never know if you'll need it and it won't hurt to have it if you don't.

If you overheat your car, it will come in handy there too.

Food

For your team or family. Ghosts don't have to eat but kids, wives or husbands and team-mates do. They all tend to become evil and ugly if they aren't being fed (except for the ghosts).

I become a bear if I don't eat. In the dark, nothing makes people laugh more than a growling stomach. Then the investigation goes to heck until control returns.

Money

For emergencies. I used to keep traveler's checks when I was carrying a large amount of money but I found that I don't have to do that now with the invention of the debit card.

A credit card for emergencies is perfect but it seems lately they all have high interest rates.

A word of financial advice: Pay the card off with-in 30 days of using it.

Calling Card

For emergency calls. Cell phones don't have "seamless coverage" and I don't carry lots of change for payphones. The calling card is a big lifesaver in these situations. Also, if your cell dies or is dropped in water etc. The calling card is all you'll have to fall back on.

First-Aid Kit

I strongly suggest you keep a few on site and one in EACH vehicle during your investigations. Also put flares and or reflectors in each vehicle for safety at night.

Uniforms

If you belong to a team or group; uniforms are really mandatory. Leave the weird hats and leather vests behind and you'll be welcomed. Would you take someone serious if they acted strange, giggled all the time and whispered? I don't think I would want someone like that in my home. I would be very uncomfortable.

Black or blue polo shirts are best for night work. I feel that big white letters on the back that say ***CREW*** are the best way to go. At night it's easy to identify a team member from someone that is an outsider. I was confident and safer when I was a cop because I could tell who the good guys were from the bad guys while running them down. Other officers could identify me from behind. Uniforms also create a professional appearance and a team work atmosphere.

The Investigation

The Nuts and Bolts of Ghost Hunting

Researching

Researching is simply going to the public library a looking up information about old buildings learning about the property in its history the reviewing old newspapers or microfiche or on the Internet now, of any terrible events, crimes or murders that occurred on the property. Talking to previous owners or property managers or residents that may know the history of the building or property would be included in the research part of your investigation. Good sources are the public library and more specifically the reference desk.

The Chamber of Commerce is a good place for information. They can usually tell you a lot about the town or city or point you to someone that knows the history of the location.

This part of the investigation should be entrusted to team members that are very thorough and determined. Researching a house or building or property should be done prior to any investigation. Knowing the history of a location will help you identify what spirits may be haunting that location and assist you in communicating with these souls.

Knowing the history will show the client how professional you and your team are.

Safety

Your safety in the safety of your team members is the most important thing during an investigation. Your team must understand that yelling and running in the dark or low light conditions is extremely dangerous.

An investigation should always be carried out with groups of two or more per group, nobody should ever be alone. Here are some things that you probably have not thought about. It is unlikely that contact with a spirit will be harmful; however it is important to know that at times, items can be thrown by an unseen force and cause injuries. Although these injuries are not life-threatening in most cases, they can really hurt and cause great discomfort. It's also unnerving to have something like this happen to you.

Another thing to keep in mind is any contact with a ghost, whether it be a touch or the movement of an object will be a life-changing event for the person affected. The stress from something like this can cause a heart attack even and so precautions need to be taken when an event like this occurs.

Before an investigation occurs team members should take the time to protect themselves from evil spirits by a saying a prayer; carrying a cross or religious item or a Bible. It is important to remember that your convictions and faith in God will protect you from evil or malevolent spirits. These spirits gain their energy and momentum from fear. The more terrorized and afraid you are the more energy you expel. If you or your team members become frightened; the safest way to handle the situation is to leave the premises immediately until you recover.

One thing team members are fearful of is that a ghost or spirit will follow them home after an investigation. As much as this is unlikely, if it does occur, a team member and their family must be very forceful and direct and order the spirit to leave their home immediately. They must command the spirit to leave their home and never return.

Protecting Yourself from Ghosts

Ghosts won't hurt you anymore than you can hurt yourself. I have never heard of anyone being physically hurt to a point that they needed serious medical attention. Spirits can move things and being hit with small objects can cause a bump or small cut. I think if you don't aggravate the spirit you probably won't have a problem.

Scratches and slaps are more common and those are few and far between if occurring. I've been ghost hunting for 35 years and never had a problem.

I say a prayer for protection and then I talk to the spirits and let them know I'm not there to kick them out or harm them in anyway. I treat the situation as if I were talking to a child or a senior citizen. With kindness and respect. Do that and you should be fine.

At the end of your investigation say a closing prayer and wish them well. Thank

them for communicating with you and if your going to be back, tell them when.
If the resident is afraid, ask the spirits to please refrain from bothering the residents and respect the home as their (the residents) home now. If requested by the property owner, command that the spirits cross-over into the light and say a prayer for them with the residents.
Empower the residents to take back their home and to demand the spirits leave at once. This seems to work with most if not all spirits. The ones that won't leave are ones a demonologist will have to deal with personally. A minister or priest can perform a house blessing and demand the spirit leave in the name of GOD. I won't go into that except to say that you will need to provide proof of an malevolent or evil spirit. Sound recordings and video are great for this and helps residents get help.
Salt also is a form of protection. The power of the salt crystals provide

protection. A circle of salt unbroken is believed to protect against evil and it works with the person standing in the middle of the circle.

Pre-investigation

This is where the planning begins. How successful the investigation is will be determined during the pre-investigation.

It's all in the planning. Big trips need planning three to six months before. If you have a team then they need to plan and have time to start saving money. If you're taking the family then motels and restroom / food stops need to be planned for.

Reservations will get you discounts if you plan far in advance and you'll have some great accommodations at cheap motel rates. Off-season travel is preferred if your going to investigate haunted hotels. Winter is the best time to hunt ghosts as they like the cool dry air. These weather conditions create static shock. When it does occur ghostly activities usually occur too.

When you arrive at the location,

you have to do a couple of things. Meet with the owner or resident and walk through the location while they tell you where the hot spots are.

Taking **baseline readings** is next. These are temperature and EMF readings through-out the entire location and specifically in the hotspots you were shown earlier. Also take note of where the power panel is and wiring and plumbing in attics and basement if possible. Be sure and take readings close to the appliances in every room. Microwave ovens, computers, electric blankets and televisions put off high EMF's.

Now set up your equipment. Make sure you arrive with plenty of time to set up in daylight. It's difficult to set up at night; even indoors. Tape down cables that cross walkways. Inevitably, someone will trip and fall in the dark. Taping cables down prevents injuries and keeps these cables from being damaged or becoming a

tripping hazard.
Now the fun! Decide on a plan of action. Send one group upstairs and the other downstairs for a certain amount of time, then reverse the teams and see what they discover afterwards. If you are in a single level send each team to opposite ends of the building.
Compare notes and go back and disprove evidence as much as possible. Whatever you can't disprove is probably paranormal activity. If you have even a doubt that it's paranormal activity then rule on the side of it not being paranormal.

Hot Spots and Cold Spots

Hot spots and cold spots are usually caused by spirits drawing the energy out of the room. Usually the temperature will drop ten to twenty degrees in the location where the spirit is manifesting. During this time your camera, flashlight batteries and radio batteries may all go dead suddenly. This is the spirit or spirits drawing energy during manifestation.

People have also reported becoming incredibly hot. Much like cold spots, hot spots also indicate a spirit or entity drawing energy. These hotspots are less common but are still a manifestation just the same. It usually feels like a fever. A digital thermometer can read these easily.

This is the time to take pictures on your instant picture or digital camera. As soon as you detect a cold spot, take a picture from different angles and afterwards write down the time, temperature and location. Record who is with you too. If one person has a lot of activity when no one else is around; your either being toyed with or your dealing with a poltergeist.

When you encounter a cold / hot spot do the following:

- **Take pictures**
- **Using your infra-red thermometer measure the temperature of the spot inside and outside; away from the spot.**
- **Record the differences.**

Photography

A few words about pictures. ORBS are not considered evidence. They can be a small insignificant piece of a larger picture but by themselves they carry no weight. Sorry. How can you prove what they are?

A picture of an apparition is great but again, there has to be photography from several angles of the same object. This provides a three-dimensional aspect to the picture and is hard as heck to replicate (fake) exactly.

I can create a great ghost picture using Adobe Photoshop. It's how I detect if a picture is a fake too. I have many graphic's programs to investigate pictures for just this reason.

Put orb pictures in a file and save them with other bits of info from your investigation. Together is where their value comes in.

Orbs are invalid for many reasons.

Here are a few:

- An insect can look like an orb on film. Unless you are using HIGH SPEED 1/2000 of a second exposure settings, there is no way to confirm it's not light reflecting off an insect, rain, dew or dust.
- Digital and regular (35mm) film can't catch more than a blur (except under the above circumstance) at lower settings in low light.

Please do not claim that light orbs are evidence. Real paranormal investigators will politely ignore you as an amateur and your credibility will take a big hit. Add them to the box of non-useable evidence.

Interviewing vs. Interrogation

Let's discuss interviewing versus interrogating. In an interview; you're asking people for information that you want them to give to you freely. This means that you're not leading them with questions that will elicit a specific answer. You want to ask open ended questions that will allow all you to gain a great deal of information. Here's an example of an **open-ended** question and a **closed ended** question. **Open ended**: Tell me what you're experiencing in this room... **Closed ended**: You mention this room is haunted, what does the ghost look like?

As you can see one question allows a complete explanation and description; the second question assumes there's a ghost already and is much more direct; forcing the resident or property owner to become defensive or implies there is actually a haunting.

You want unbiased information only. Don't poison the well with a loaded question. Remember, you are there to investigate the possibility of a haunting. You can't confirm a haunting before you investigate – right?

An interrogation is much more forceful. During an interrogation; the person you would be talking to would be intentionally withholding information and you as the interrogator would be trying to force an answer from this person under duress.

It is extremely aggressive in its approach. One thought that comes to my mind is that of an SS officer asking questions and slapping the interviewee until the interrogator hears what he wants to hear. Quite obviously this is a totally unacceptable way to behave during an investigation. Unless you want to stay in jail for a while; I wouldn't suggest it.

After the Investigation

After your investigation is complete there are a few things you need to do. Walk-through and collect all your cables and replace all furniture that may have been moved. Let the homeowner or property owner know that your investigation is complete.

Do not discuss any phenomena or occurrences with the homeowner until you have reviewed your video, photographs and audiotapes and reviewed log sheets from each team member. The best way to do this is to tell of the owner or resident you'll contact them in a few days. Also let them know that they can contact you at any time before then if anything else occurs.

If you suspect a ghost or entity at the home or property, you can empower the resident or owner of the property on how to demand that the ghost leave. Frequently; much phenomena is occurring from natural things occurring in the house, and can be explained. During an actual haunting the spirit is gaining energy from the fear of the resident or people in the location of the property. The more energy; the more interaction and the more interaction the more fear and of course the more fear the more energy.

The resident has to understand that they do have the power to take back their home and make the spirit or ghost leave. This will work with an entity whereas it won't work with an imprint. The reason this won't work with an imprint is because an imprint is much like a video recording playing over and over again under the right circumstances. The good news here is that an imprint won't harm you at all. An entity can. Usually they don't though. It's very rare.

Usually people want to know if they are going crazy or if there really is a ghost. Once confirmed, you'll find the people relieved. Be sure to explain that you don't perform exorcisms. You're a ghost hunter not exterminator. Unless you have a lot of experience in this and strong faith, it's better not to mess with a malevolent spirit or entity. Definitely empower the owner or residents to demand the spirit leave and if you want; you can say a prayer with them.

I would suggest finding a demonologist to rid a malevolent spirit. Ministers and priests can help here if you bring them a lot of evidence. They take it very seriously and you have to really prove to them something is going on. Otherwise they'll have no interest in assisting you and your credibility will be lost with them.

Log Books

A log book is just that, you're logging information about the haunting or your investigation. These are your notes. The log book should be kept by the team leader. Each team member should fill out a piece of paper telling in detail what experiences they had throughout the evening and the approximate times they should write these down on a little notebook that they keep in their pocket for example:

- At 10:03 p.m. in the upstairs east bedroom, I experienced cold spots throughout the room measuring 52°. The air temperature in the room is normally 75°.
- At this time Mark reported the hair on the back of his neck standing straight up and a moment later we saw what looked like orbs.
- EVP and photographs taken.

At the end of the investigation everybody turns in their little log books and the main log book records all this data rewritten again. It is this log book that you will use when you speak to the resident about the phenomena they are experiencing.

Reviewing Data

This of course is simply reviewing all the information from a log books, deal, photographs, and recordings. This is the most time consuming more of an investigation in the work is tedious. Reviewing 40 hours of videotape will wear you out. It's important to remember you may only have a few seconds of evidence on this forty hours of tape and inattentiveness will cause you to miss that.

Audio recordings again are the same inasmuch as you need a quiet room and a lot of patience while reviewing an audiotape or a digital recording. Most ghost hunters / investigators use a computer program and download from a digital recorder into their computer so they can visually see the soundtrack in addition to hearing it. There are many good programs out there that you can purchase for this, see our tools page for more information about the software.

The AAEVP has an excellent site about EVP collection. Visit it at

www.aaevp.com . They also have a free program to open, view and manipulate the sound recording on your computer.

It's an excellent site to learn about and discuss EVP collection.

Disproving Evidence

Let's discuss evidence for a moment. For investigators, evidence is everything; but what is evidence exactly. A photograph of a disembodied spirit or an apparition is fantastic evidence but unless you have two different cameras with two different angles of the same photograph; the evidence is inconclusive.

Let me explain why: a photograph can be faked. Two photographs of the same apparition at the same time will give you two points of view; this is better known as three dimensions. With these two separate pictures a photo lab can compare both photographs against each other scientifically to determine if the phenomena - in this case the disembodied spirit - has been faked or is real. This is solid evidence because it would be very difficult to fake the pictures. It would be nearly impossible. Of course it can be done but it would be very difficult to achieve exacting results. Of course a single photograph can be quite compelling also but not significant evidence.

Deciding to keep evidence or throw it out is a hard part of any investigators job. Take for example light orbs. A light orb can be an entity or spirit manifesting itself. It can also be mist, dust or even an insect. On film or video; is virtually impossible to tell one from the other. Although a light orb indicates an existence of something; it's inconclusive as evidence. It does add to the evidence but only in a minor way. Team members may think they are being touched or feel cold spots and although compelling to the person affected by these things; these are nothing more than feelings and therefore are not considered hard evidence either.

Keep this in mind, anything that you question should be thrown out and anything that you can say with 100% absolute certainty is proof of a haunting; should be kept. This means that about 99.99% of your evidence will have little or no value. The best thing to do with this type of information is to file it away in your file cabinet in a folder with the locations address and contact information for future investigations.

Client Reveal

The client reveal is the meeting you schedule with a resident or owner of the property. Is during this time that you present your evidence and let them see for themselves what you found. You should never declare a location as haunted unless you're absolutely sure it is.

This is also where you tell the owner or resident about environmental factors such as high EMF readings, knocking pipes or other natural phenomena that may appear to be paranormal in nature.

At the end of your reveal; you should ask the resident or owner how they feel about your evidence and offer to help if you feel that's what they want. This is the time that you would empower them to take back their home or property and teach them to our where demand the spirits leave the property immediately and go into light.

It's OK to let them decide for themselves if their property is haunted or not. Don't jump the gun… let them come to their own conclusions.

Declaring a house or property haunted could cause the property to lose value if it's up for sale. Most states have disclosure laws and technically, the home owner must disclose everything about the property; even *GHOSTS*.

Follow-Up

A follow-up is simply calling back to see if the property owner or resident is having any problems and just checking up on them. This is let them know that somebody cares about them and is still there. Most people feel completely helpless during a haunting and there is great comfort in knowing that they are not alone. It helps a lot that people don't think they're crazy. It's also a very professional way to close an investigation.

Links Page

These links are to great websites that I'm sure you'll enjoy.

This is my website. If you have questions you can email me or write articles for The Paranormal Investigator quarterly newsletter:
http://www.the-paranormal-investigator.com/

Sign up for our quarterly E-zine
http://www.the-paranormal-investigator.com/tpimailform.html

The American Association of Electronic Voice Phenomena:
www.aaevp.com

Ghost Tech software: This is freeware for ghost hunters with lots of utilities. Great deal for the price!
http://www.download.com/Ghost-Tech-Paranormal-Investigator/3000-

2054_4-10556320.html?tag=pdp_prod

The Atlantic Paranormal Society: Excellent info resource.

TAPS is a non-profit organization dedicated to educating and helping those interested in or troubled by ghostly phenomena.

http://www.the-atlantic-paranormal-society.com/

Equipment and Software Resources

I'm not rich so I search for great deals everywhere eBay included. My wife and budget demand I search high and low for deals so here are the best sites to get info, equipment and software from. If you find any other sites, email me with the web address so I can put it on my site.

SUPERMETERS STORE

has some of the lowest prices I've seen anywhere and they carry every piece of equipment you could ever need. Call Richard Fitton and tell him I sent you. Here's their link:
http://www.supermeters.myzen.co.uk/store/

Ghost Tech created a freeware utility program especially for Ghost Hunter's needs and distributes it on Download.com Here's a generic link and search for "**Ghost**

Tech Paranormal Investigator".
http://www.download.com/2001-20 4-0.html?tag=hdrgif
Or this direct link to the download site.
http://www.download.com/Ghost-Tech-Paranormal-Investigator/3000-2054 4-10556320.html?tag=lst-0-1

Audacity Here is a freeware sound program for EVP work. It is great for the price. It even includes a white noise generator and is easy to use.
http://audacity.sourceforge.net/

The **AAEVP** has a lot of useful information about EVP collection and a support circle of EVP enthusiasts
http://www.aaevp.com/

Epilogue

I truly hope to hear from you as you embark on your scientific journey. I have taught you all that I can and from here you'll have to go out on your own. Just don't go alone. It's more fun with a couple of people and from a safety point of view always travel in pairs.

I wish you the best and that you go with your GOD on your journey of enlightenment.

One last note: Always feel free to contact me with your questions and write me about your investigations. I love to hear about others adventures. That's why I wrote the book. I'll post your stories on my website if you don't object so don't be shy. It may take a while but I always answer my email. Have fun and I hope you enjoy being scared a lot, it's a real rush.

Mike Atencio

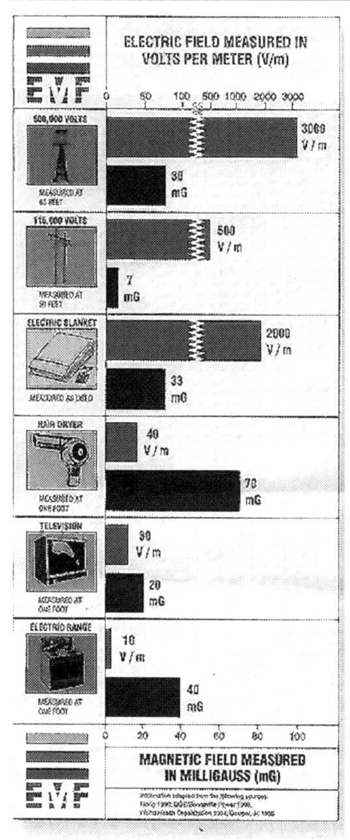

Example of Dept of Energy EMF Guide

This guide shows how EMF's can affect you and your equipment. You can get one from **The Department of Energy**

A reading of 2 to 8 milligauss is in the range of ghostly activity or phenomena. Above this is some type of electrical field. High field readings can be unshielded power panels, receptacles, appliances etc.

Very high readings are dangerous and can cause *illness, dread, dizziness, headaches* and *hallucinations.* These should be repaired immediately by a professional electrician.

Here's a simple test: Take your meter and use it around the microwave oven while it's off and while it's on. You can see that they will *cook* you with-in a foot of the oven

Creating a Ghost Hunter's Club

I believe it is time for ghost hunter's to start charging for their services. I don't like to work for free and I don't think anyone else should either. The real question then becomes how much should you charge? Well how many people do you have with you? Figure minimum wage for each member of your team and about $10.00 to $20.00 an hour for yourself.

Money should really go back into your "Kitty" for new equipment and travel expenses. Of course, volunteer's are easy to find and the economy in your area may make it so that people can't afford much if anything at all. A sliding scale could be created so all can benefit from your services. Ask for donations.

Insurance is a must if you charge for services and a business license is required as well. This means you probably want to become incorporated. Non profit organizations can be set up but there are a lot of headaches that go with this form of

business. An LLC may be the way to go. Get an attorney and ask him or her what each has to offer.
Privacy is a major issue. You cannot release any pictures or personal information without written permission. You can be sued for libel. Discretion is mandatory.
You can not call yourself a parapsychologist. This is against the law because you do not have a doctorate degree in psychology. Paranormal Investigator is the key term here.
Your team needs to ELECT a President, V.P., Secretary and Treasurer if you're incorporated. That means you can be voted out of office if you fail in your work.
If you charge, you must provide the service contracted for. Set-up and monitoring of a location with exact records of who did work and how long they did the work. A signed contract is required by the owner of the property, not the resident. You can always opt to do a pro-bono (donate services) and write them off at the end of

the fiscal year. You also have to figure depreciation for your equipment and vehicles as part of your business.

Get an accountant to help you with this. An accountant and attorney will earn their money and you'll save more money than you'll spend with their services.

I want to stress that there are many scam artists posing as "mediums, psychics and palm readers" out there.

You must strive to be as honest as possible in every investigation you are a part of. This means that you need to do a background and fingerprinting of each member on your team. It costs about $15.00 for the fingerprinting and can be done through your local police department or Sheriff's office. The criminal background check is usually about $10.00 to $25.00 for each. You can do the personal background check yourself.

Notes:

Notes:

Notes:

Notes:

Notes: